Einstein Simplified

RUTGERS UNIVERSITY PRESS
New Brunswick, New Jersey

Einstein Simplified

Cartoons on Science by Sidney Harris

Fifth Printing, 1997

Library of Congress Cataloging-in-Publication Data

Harris, Sidney.
Einstein simplified : cartoons on science / by Sidney Harris.
p. cm.
ISBN 0-8135-1410-X (pbk.)
1. Science—Caricatures and cartoons. 2. American wit and humor, Pictorial. I. Title.
NC1429.H33315A4 1989
741.5'973—dc19 88-31284
CIP

British Cataloging-in-Publication information available

Manufactured in the United States of America

Einstein Simplified

"THERE GOES ARCHIMEDES WITH HIS CONFOUNDED LEVER AGAIN"

"JUST CHECKING."

THIS IS A RECORDING.

DOESN'T BOTHER ME. I'M A HOLOGRAM.

"WE COLLABORATE. I'M AN EXPERT, BUT NOT AN AUTHORITY, AND DR. GELBIS IS AN AUTHORITY, BUT NOT AN EXPERT."

ENTERING
CALIFORNIA
ENTERING
SAN ANDREAS
FAULT
S. Harris

"IT LOOKS LIKE FRED HAS GONE SOLAR."

s.harris
— MESOZOIC ERA → | ← CENOZOIC ERA –

"IT FIGURES. IF THERE'S ARTIFICIAL INTELLIGENCE, THERE'S BOUND TO BE SOME ARTIFICIAL STUPIDITY."

ADULT BOOKS
ALL NEW VIDEOS
YOU MUST BE OVER 18
Set Theory
MONOCLONAL ANTIBODIES
PALEO GEOLOGY
Z PARTICLES
Z PARTICLES
Z PARTICLES
S. Harris

RIBS.
S. Harris

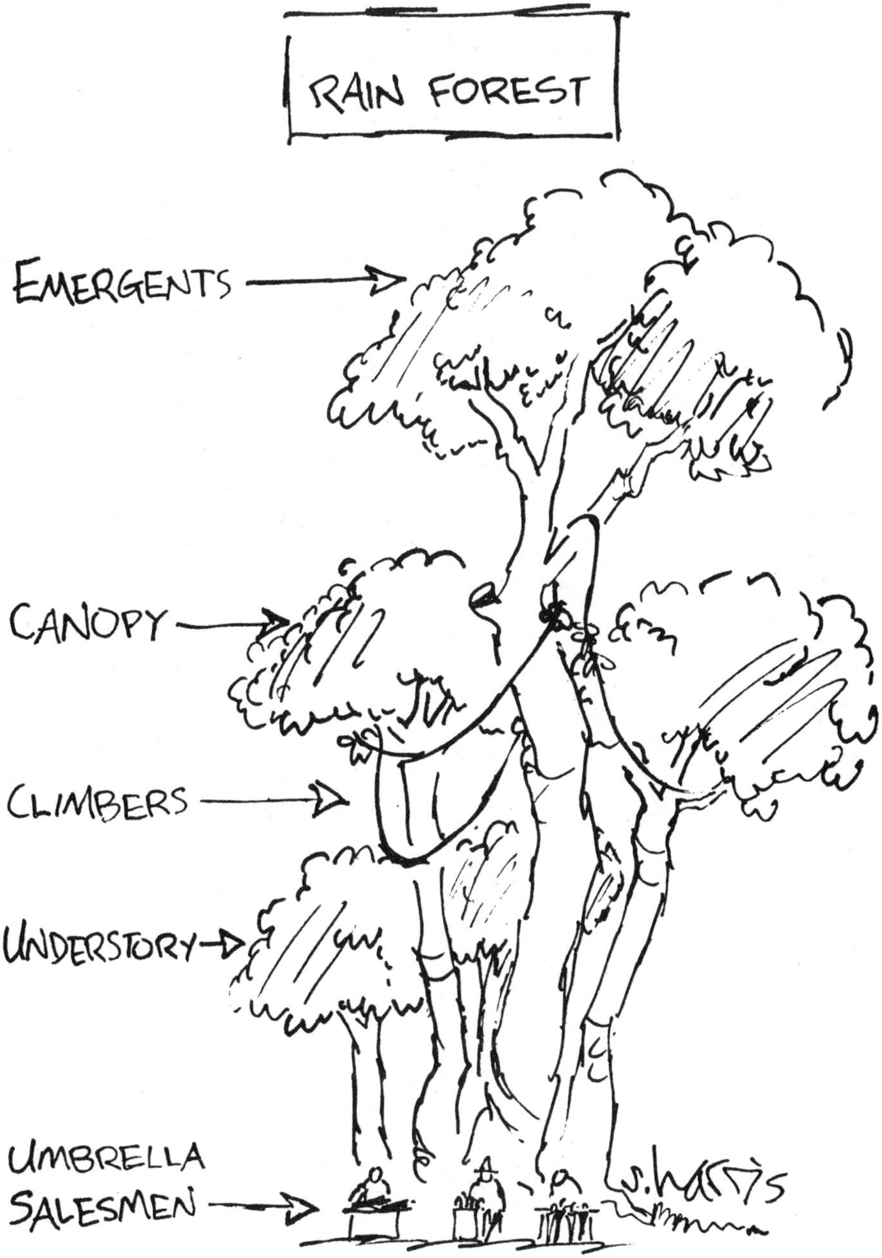

RAIN FOREST

EMERGENTS

CANOPY

CLIMBERS

UNDERSTORY

UMBRELLA SALESMEN

S. HARRIS

"THE ENVIRONMENT PEOPLE ONLY WORRY ABOUT ENDANGERED SPECIES, NOT ENDANGERED INDIVIDUALS."

STANDUP CHEMIST
...SO HE COMES BACK INTO THE ROOM, HE TURNS AROUND, HE PUTS THE CHICKEN ON THE TABLE AND HE SAYS, "ALL RIGHT, POTASSIUM CHLORIDE."
EXIT

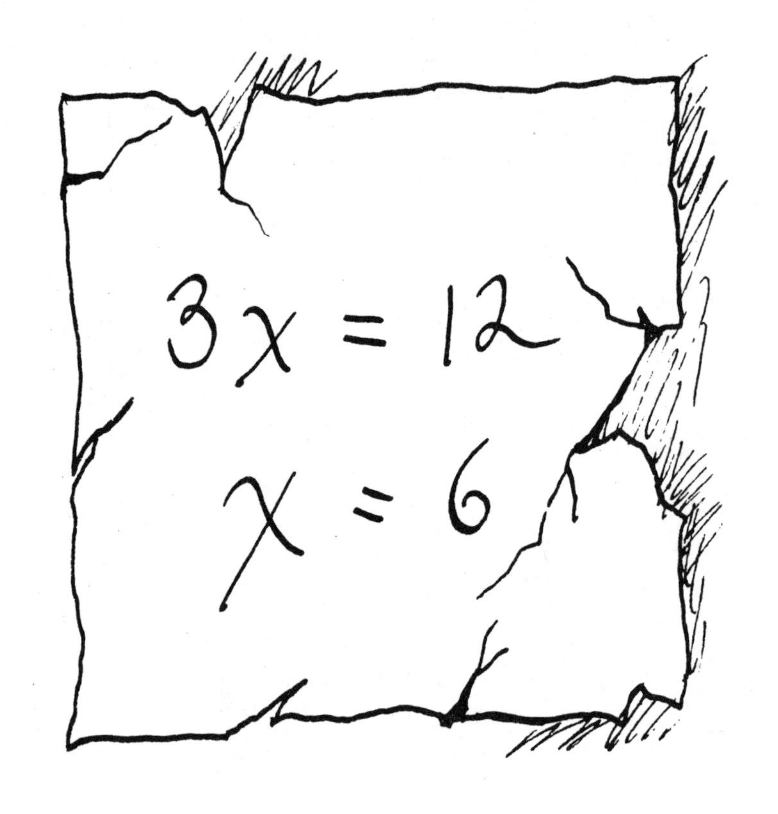

EINSTEIN'S
FIRST EQUATION

WHEN SCIENCE CATCHES UP TO SPORTS
I WANT $400,000 A YEAR, A BONUS FOR EACH STEP THAT LEADS TO A NEW PRODUCT AND A FIVE-YEAR CONTRACT. OR TRADE ME TO ANOTHER LAB.

I think I've made a breakthrough concerning the structure of matter.
S. Harris

"IF I LEARNED ONE THING IN MY LONG REIGN, IT'S THAT HEAT RISES."

THE LAST BRONTOSAURUS ON EARTH UNABLE TO UNDERSTAND WHY HE CAN'T GET A DATE FOR SATURDAY NIGHT

SCHOOL OF OLOGY
ANTHROP 301
ARCHAE 126
BACTERI 109
BI 326
ENTOM 217
ETYM 221
GE 204
PALEONT 113
PHYSI 312
PSYCH 209
TOXIC 307

"SINCE THAT YOUNG DARWIN FELLOW WAS HERE, THEIR SPECIES CERTAINLY HAS EVOLVED."

MATTER
ANTI-MATTER
S. Harris

"UNFORTUNATELY THIS LAB IS FUNDED ONLY BY AS MUCH GOLD AS WE CAN MAKE FROM LEAD."

TECHNOLOGY INC.
NOT HIGH TECH
NOT LOW TECH
BUT JUST THE RIGHT TECH
S.Harris

1.
2.
3.
4.
GUN POWDER
S. Harris

"DON'T FORGET—KEEP THE POTASSIUM CHLORIDE IN A SEPARATE CONTAINER."

"So, by a vote of 8 to 2 we have decided to skip the Industrial Revolution completely, and go right into the Electronic Age."

"FORGET ENLIGHTENMENT. I WANT YOU TO CONCENTRATE ON THE STRUCTURE OF THE PROTEIN MOLECULE."

TWO CREATURES, AFTER ATTEMPTING TO CATCH THE SAME INSECT, NOW JOINED TOGETHER BY THEIR STICKY TONGUES.

BUREAU OF STATISTICS
OFFICE HOURS:
S. Harris

I'VE DONE IT— I'VE FOUND THE MOST BASIC PARTICLE!
I'VE FOUND THE PARTICLE THAT MAKES UP THAT BASIC PARTICLE!
I'VE FOUND THE PARTICLES THAT MAKE UP THE PARTICLES...
S. Harris

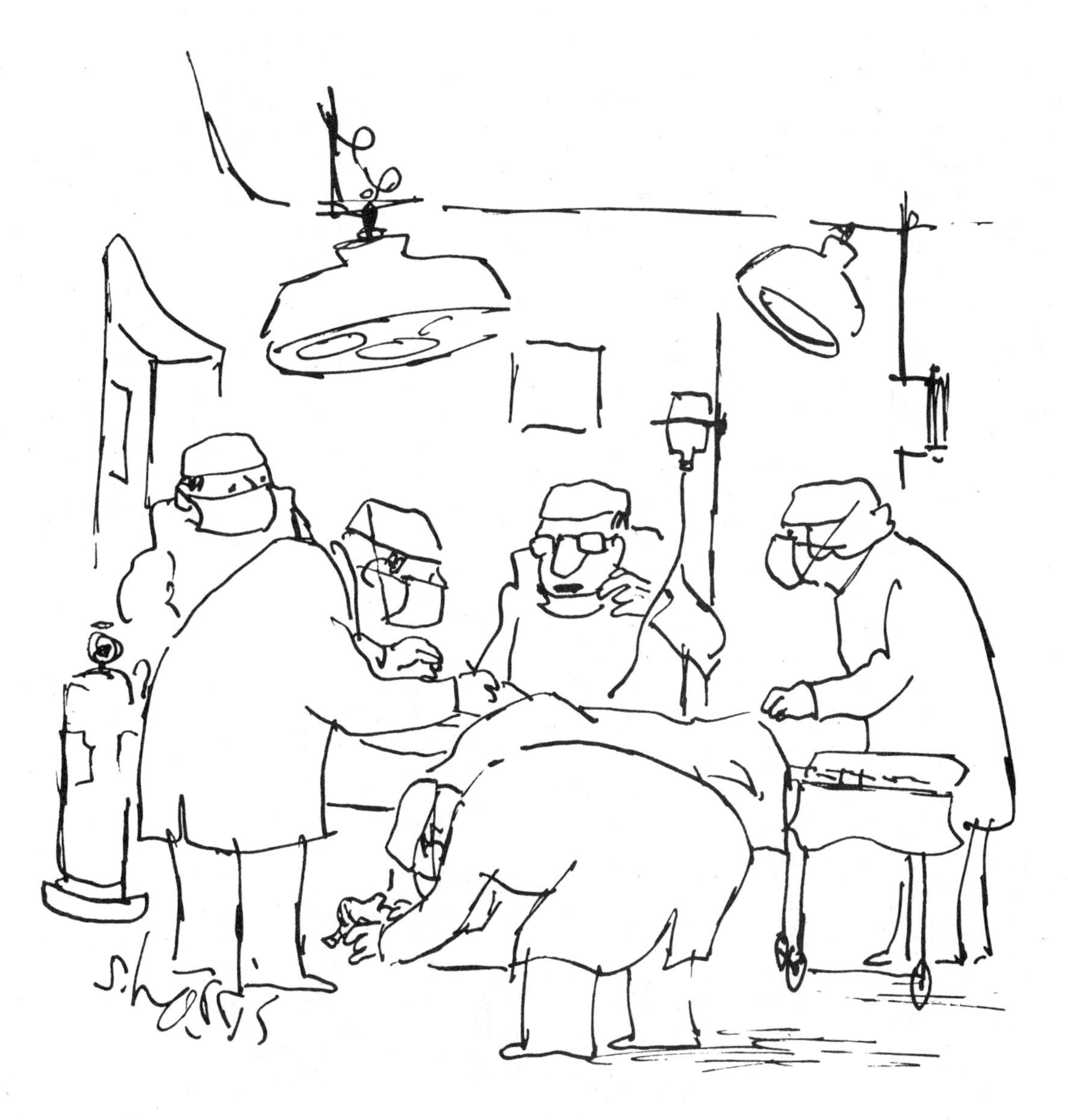

"ALL RIGHT, SO HE DROPPED THE HEART. THE FLOOR IS CLEAN."

IMMEDIATELY AFTER ORVILLE WRIGHT'S HISTORIC 12-SECOND FLIGHT, HIS LUGGAGE COULD NOT BE LOCATED.

"THEN, AS YOU CAN SEE, WE GIVE THEM SOME MULTIPLE CHOICE TESTS."

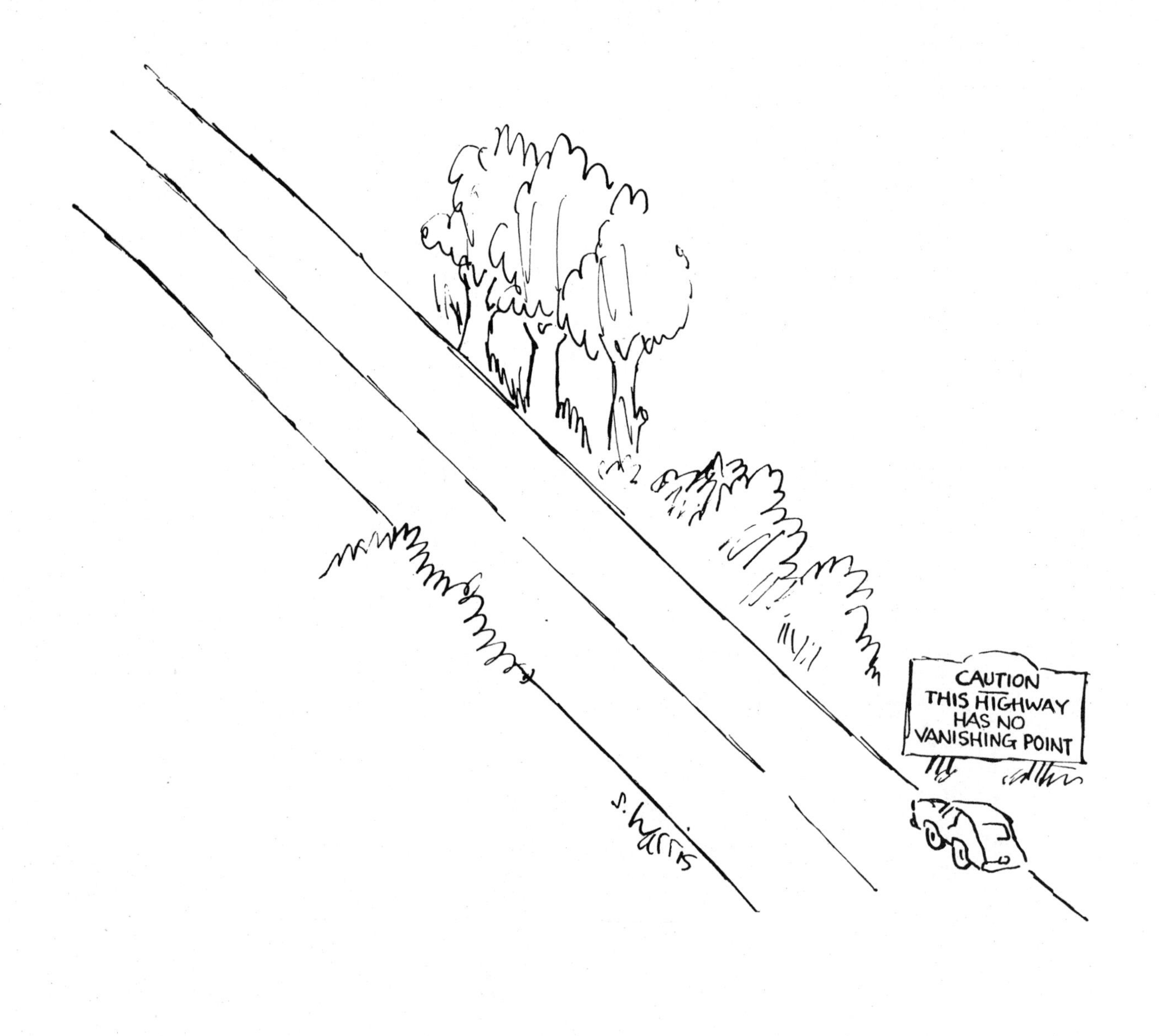
CAUTION
THIS HIGHWAY
HAS NO
VANISHING POINT
S. Harris

"IT'S AN INTERESTING PSYCHOLOGICAL PHENOMENON. THEY THINK HE'S THEIR MOTHER. SO DOES HE."

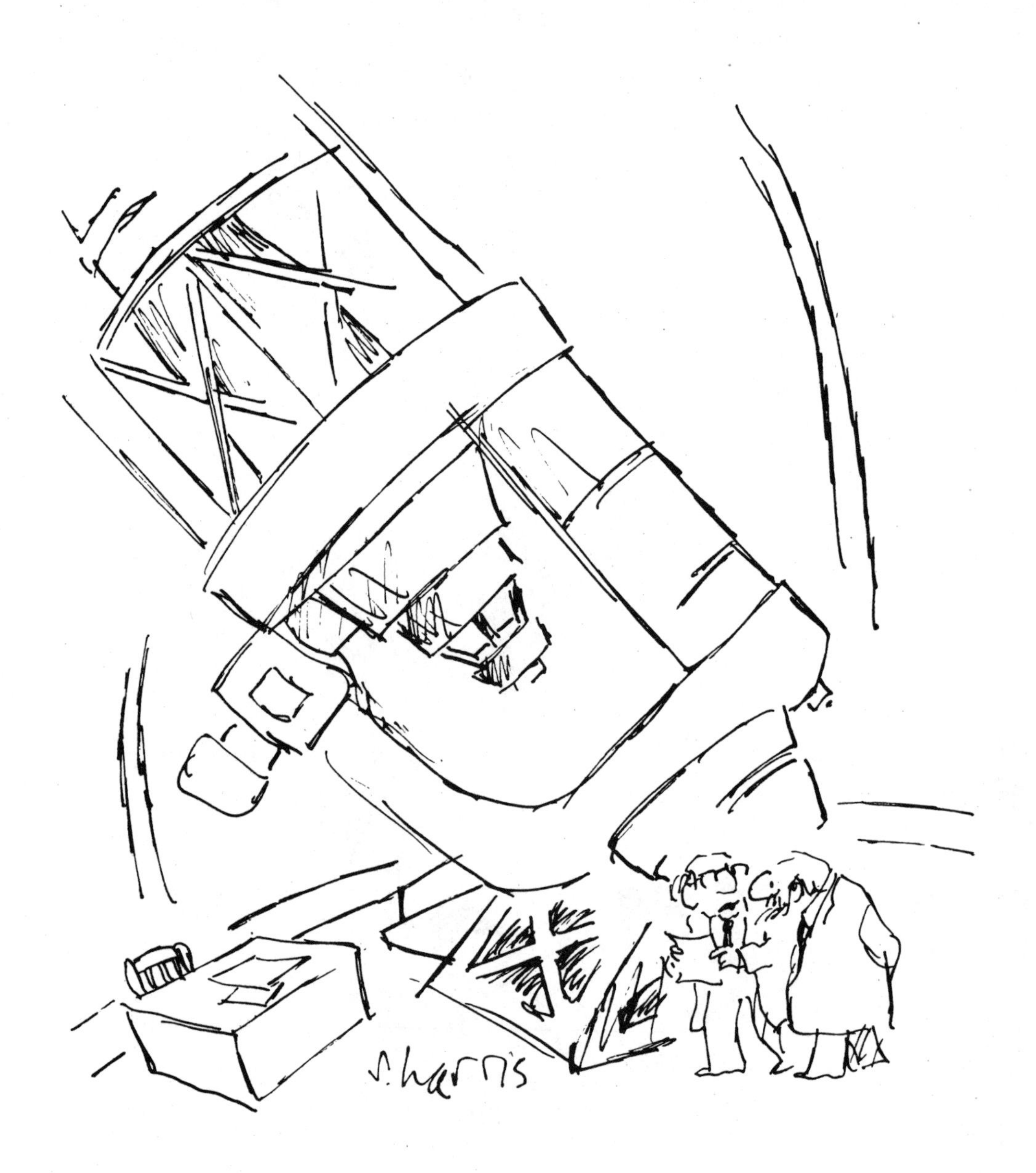

"THE ONLY PART OF THE UNIVERSE WHICH ISN'T EXPANDING IS THE BUDGET FOR THIS PLACE."

S. Harris

TRYING TO DESCRIBE THE SIZE OF THE BIG BANG

INSTITUTE FOR
NONVERBAL COMMUNICATION
S. Harris

How's the SPLEEN?
The SPLEEN? Where is it?
I never studied the spleen.
I was left back in my spleen class.
I FAILED spleen.
I thought we didn't have to know the spleen.
S. Harris

"BIGGEST DAMN VIRUS I'VE EVER SEEN!"

HOW TO PLAY WITH A RADIOACTIVE BASEBALL

GREATLY ENLARGED VIEW OF
NEWLY DISCOVERED PARTICLE

BROTHERS AND SISTERS, AT THE TIME OF 10^{-33} SECONDS AFTER THE BIG BANG, THE HEAT WAS ENORMOUS.
VERILY, IT WAS OVER 10^{32} DEGREES!
MATTER AND ANTI-MATTER AROSE!
HALLELUJAH- THEY ANNIHILATED EACH OTHER.
AND THE UNIVERSE WAS FILLED WITH PARTICLES...
AMEN! QUARKS
AND GLUONS.
YEA, LEPTONS
BELIEVE
S. harris

NO ONE IS LOOKING. I'LL JUST ADD A LITTLE MERCURY NITRATE.
No one is looking. I'll just add a little mercury nitrate.
NO ONE IS LOOKING. I'LL JUST ADD A LITTLE MERCURY NITRATE.
No one is looking. I'll just add a little mercury nitrate.
S. HARRIS

"IT MAY NOT BE A PERFECT WHEEL, BUT IT'S A STATE-OF-THE-ART WHEEL."

TECHNOLOGY BREAKTHROUGH: THE CORDLESS TOASTER

"DON'T KID YOURSELF. IF YOU'RE TOO UGLY TO BE COOKED, THEY'LL GRIND YOU UP INTO ONE OF THOSE FISH PROTEIN CONCENTRATES."

CHAPTER I
Call me Euclid.

S.Harris

"I MISS THE GOOD OLD DAYS WHEN ALL WE HAD TO WORRY ABOUT WAS NOUNS AND VERBS."

AS SMART AS HE WAS, ALBERT EINSTEIN COULD NOT FIGURE OUT HOW TO HANDLE THOSE TRICKY BOUNCES AT THIRD BASE.

BOOLE ORDERS LUNCH

"THIS IS TERRIBLE. IF THE PLANKTON GO OUT ON STRIKE, IT'LL DISRUPT THE ENTIRE FOOD CHAIN."

INSTITUTE of
SOLID GEOMETRY
Institute of
Plane
Geometry

PEMBROOK, THE RENEGADE THIRD WRIGHT BROTHER, AT WORK ON THE FIRST ANTI-AIRCRAFT GUN

"EVERY DAY YOU SHOULD EAT SOMETHING FROM EACH OF THE FIVE BASIC FOOD GROUPS: FRIED BLUBBER, BOILED BLUBBER, STEWED BLUBBER, BAKED BLUBBER AND RAW BLUBBER."

DEPT. OF
HEALTH
GENETIC
ENGINEERING
OH-OH!
DEPT. OF
DEFENSE
BIOLOGICAL
WARFARE
S. Harris

"ALL I CAN TELL YOU IS THEY'RE NOT THE CUTE LITTLE BIRDS WHO USUALLY RIDE AROUND ON US."

"IT STARTED WITH JUST THE PARTICLES BEING ACCELERATED, BUT NOW EVERYTHING AROUND HERE HAS SPEEDED UP."

"IF ONLY HE COULD THINK IN ABSTRACT TERMS."

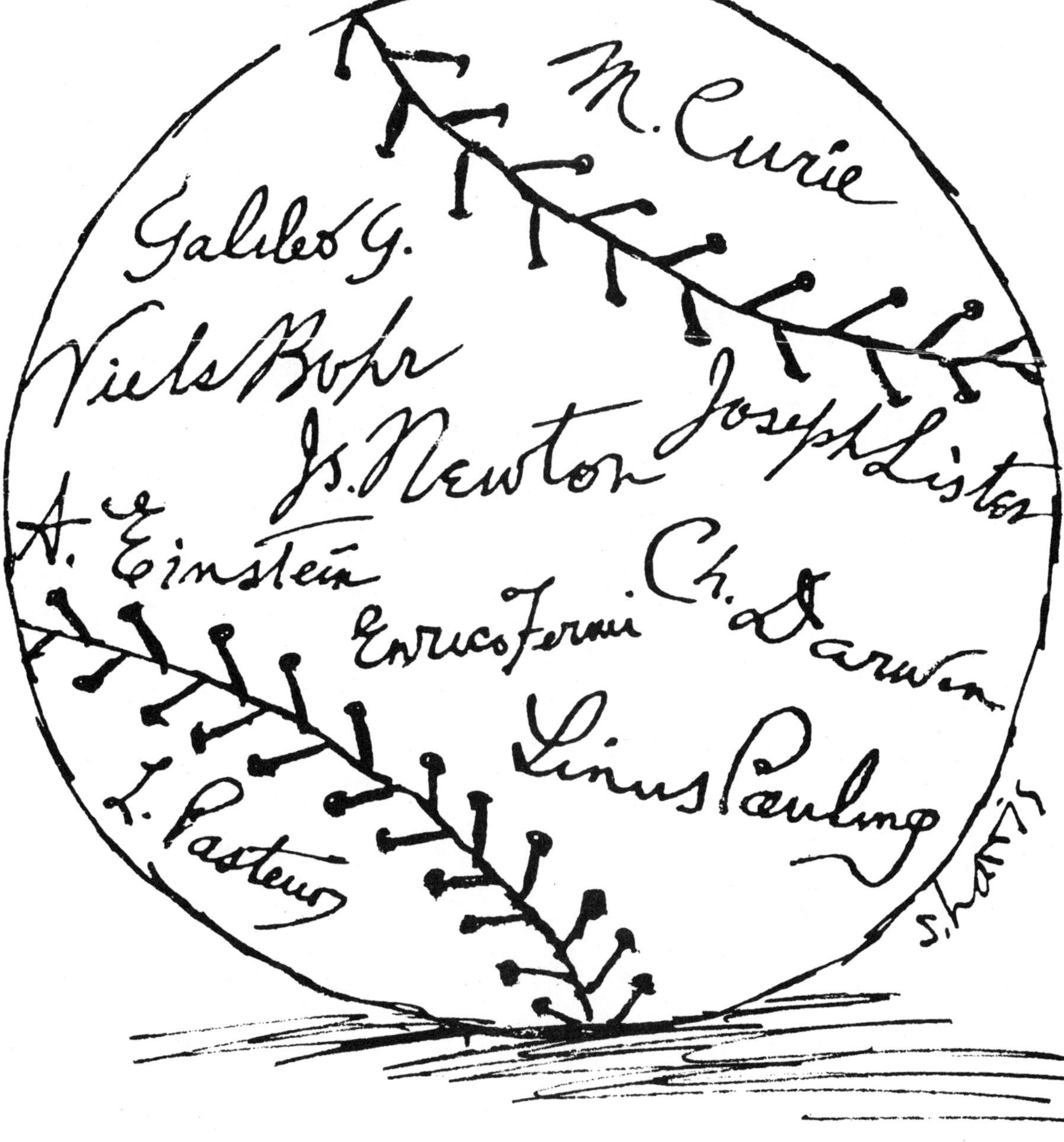

M. Curie
Galileo G.
Niels Bohr
Js. Newton
Joseph Lister
A. Einstein
Ch. Darwin
Enrico Fermi
Linus Pauling
L. Pasteur
S. Harris

"I ADMIRE THE INQUIRING MIND AND THE PRAGMATIC MIND, BUT I ALSO ADMIRE SOMEONE WHO CAN HIT."

"QUARKS. NEUTRINOS. MESONS. ALL THOSE DAMN PARTICLES YOU CAN'T SEE. THAT'S WHAT DROVE ME TO DRINK. BUT NOW I CAN SEE THEM!"

"OF COURSE IT'S SAFE. IT HAS NO PRESERVATIVES, NO ADDITIVES, NO ARTIFICIAL COLORING..."

A NATION OF EINSTEINS

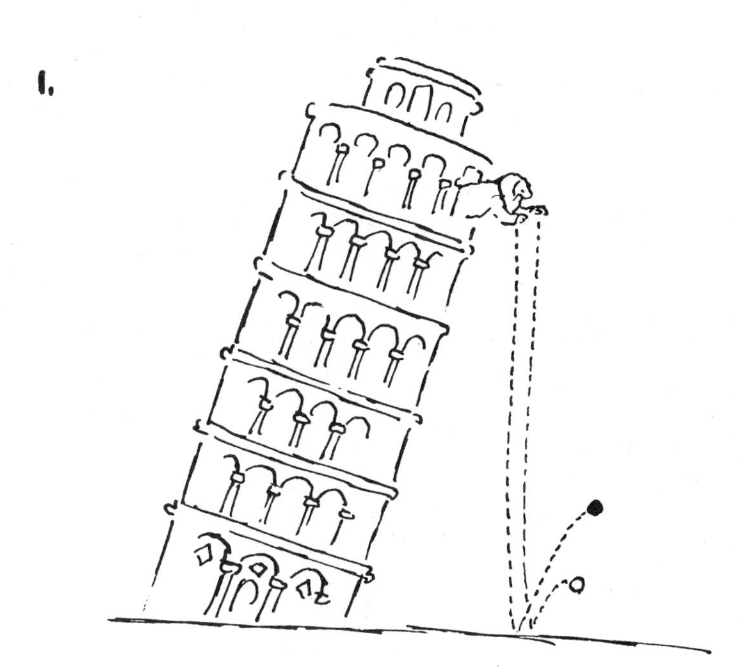
1.

2.
High-
Bouncers
12 Lire
S. Harris

WARNING
PREMISES PROTECTED
BY WIDE-RANGE
MOTION-DETECTION
AUDIO-MONITORING
SYSTEM
S. Harris

I'd like something with a high amount of vegetable protein, a bit of glucose or dextrose for energy, and some fiber.
NATURE'S WAY
One peanut butter and jelly on toast.
SALADS
S. Harris

Apples
s.harris

"A GIANT KILLER MACROPHAGE! DOCTOR, THIS IS MADNESS!"

SMOKING
NON-SMOKING

HIGH-TECH HAIRCUT

EVOLUTION UPDATE — 101,989 A.D.

WILHELM RÖNTGEN'S FIRST ATTEMPT AT X-RAYS: SHINING A BRIGHT LIGHT THROUGH MADAME RÖNTGEN

"RORSCHACH! WHAT'S TO BECOME OF YOU?"

"THE FUNNY THING IS, OUR WASTES ARE PERFECTLY HARMLESS."

So you created everything.
Yes.
Including black holes.
Yes.
Which will eventually swallow up everything.
Yes.
Including You.

I'm working on that.

S. Harris

1. It is best to migrate in
- January
- October
- February

2. Which has the most protein?
- Grass
- Seeds
- Worms

3. You are related to a
- Moose
- Owl
- Snake

S. Harris

I.Q. TEST FOR BIRDS

"M. PASTEUR, I'D LIKE YOU TO MEET SOMEONE WHO HAS ANOTHER IDEA ABOUT IMPROVING THE QUALITY OF MILK, M. HOMOGEN."

THE IRON AGE: DAY ONE

"PARTICLES, PARTICLES, PARTICLES."

1, 2...MANY
S. Harris

BARTENDER WITH PH.D. IN PHYSICS

GENETIC
ENGINEERING
INC.
A
WE-WILL-MAKE-YOU-AN-EQUAL
EMPLOYER
S. Harris

"HE SALIVATES."

Ah, ha—just as I expected.
Ah, ha—just as I expected.
Ah, ha—just as I...
S. Harris

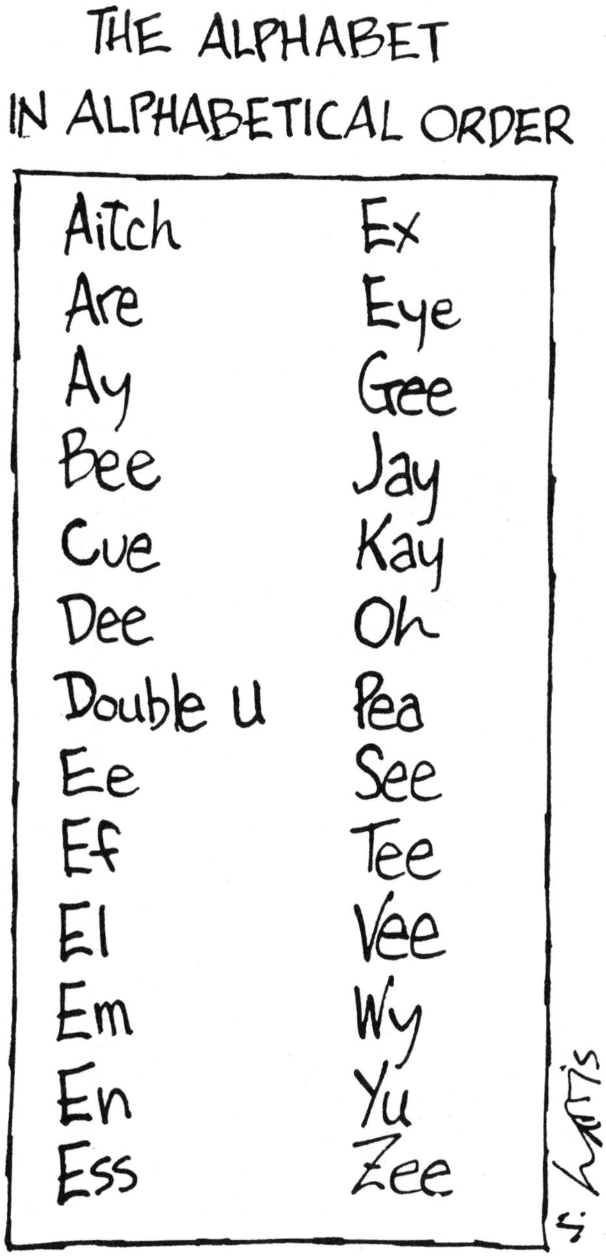

THE ALPHABET IN ALPHABETICAL ORDER

Aitch
Are
Ay
Bee
Cue
Dee
Double u
Ee
Ef
El
Em
En
Ess
Ex
Eye
Gee
Jay
Kay
Oh
Pea
See
Tee
Vee
Wy
Yu
Zee

ANTHROPOMORPHISM
LABORATORY

"ALL WE WANT IS SOMETHING NEW THAT WILL INCAPACITATE THE ENEMY WITHOUT GIVING US A BAD PRESS."

SCIENCE AND SOCIETY—1923

SCHRÖDINGER, PAULI, HEISENBERG, PLANCK, BOHR, CURIE, EINSTEIN, KIKI AND BUD VANDERVELT

"IT'S OUR NEW ASSEMBLY LINE. WHEN THE PERSON AT THE END OF THE LINE HAS AN IDEA, HE PUTS IT ON THE CONVEYOR BELT, AND AS IT PASSES EACH OF US, WE MULL IT OVER AND TRY TO ADD TO IT."

DEPT. OF ENTROPY
1967
S. Harris

"LOOK—UP IN THE SKY! IT'S A FLOCK OF BIRDS. IT'S A CLOUD. IT'S THE MONTGOLFIER BROTHERS."

EVEN AMONG THE SOCIAL INSECTS, SOME INDIVIDUALS ARE CLEARLY ANTI-SOCIAL

All cats have four legs.
I have four legs.
Therefore, I am a cat.

S. Harris

EINSTEIN SIMPLIFIED

"GRANTED, WE HAVE TO DO THE RESEARCH. AND WE CAN DO SOME RESEARCH ON THE RESEARCH. BUT I DON'T THINK WE SHOULD GET INVOLVED IN RESEARCH ON RESEARCH ON RESEARCH."

BRAINS

JAVA MAN NEANDERTHAL MAN MODERN MAN

"AS I MENTIONED NEXT WEEK
IN MY TALK ON REVERSIBLE TIME..."

"COMRADE — THE COMMISSAR OF MATHEMATICS WANTS IT TO EQUAL 29.86."

I am a think tank,
therefore I am.
S. Harris

"MY GOODNESS, IT'S 12:15:0936420175. TIME FOR LUNCH."

"LET'S GO OVER TO CELSIUS'S PLACE. I HEAR IT'S ONLY 36° OVER THERE."

PRESERVE
THESE
WETLANDS
S. Harris

METRIC CLOCK

"IN THE BEGINNING, BIOPHASE GENETICS, INC. CREATED US."

"THE RIGHT SIDE OF MY BRAIN SAYS YES, BUT I'M WAITING TO HEAR FROM THE LEFT SIDE OF MY BRAIN."

"I TRIED IT, BUT THE BLOOD RUSHED TO MY FEET."

"THE BEAUTY OF THIS IS THAT IT IS ONLY OF THEORETICAL IMPORTANCE, AND THERE IS NO WAY IT CAN BE OF ANY PRACTICAL USE WHATSOEVER."

CONFERENCE ON THE CONTROL OF TIME AND SPACE
Left to Right: ISAAC NEWTON, ARISTOTLE, H.G. WELLS, PTOLEMY, ISAAC ASIMOV, ALBERT EINSTEIN, GALILEO, COPERNICUS

"I'll be working on the largest and smallest objects in the universe—superclusters and neutrinos. I'd like you to handle everything in between."

RESEARCH
BOTTLENECKS
DEVELOPMENT
S. Harris

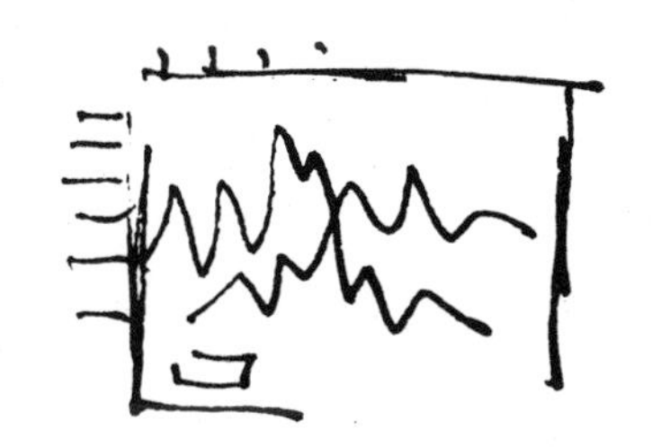

TEST
CONTROL
LUNCH
S. Harris

"WE TRIED HUNTING AND WE TRIED GATHERING, BUT NOW WE USUALLY EAT OUT."

BAN THE BOMB
BAN NERVE GAS
BAN GERM WARFARE
BAN THE DISINTEGRATOR RAY

1.

DORVAL'S
Cake Mix
INGREDIENTS:
SUGAR, FLOUR,
SHORTENING, SALT,
ARTIFICIAL FLAVORING,
BHA and BHT,
PROPYL GALLATE,
SODIUM PROPIONATE,
MONO and
DIGLYCERIDES

2.

DORVAL'S
CaKe
Mix
INGREDIENTS:
SUGAR, FLOUR,
SHORTENING, SALT,
ARTIFICIAL FLAVORING,
BHA and BHT,
PROPYL GALLATE,
SODIUM PROPIONATE,
MONO and
DIGLYCERIDES

3.

DORVAL'S
CaKe
Mix
INGREDIENTS:
SUGAR, FLOUR,
SHORTENING, SALT,
ARTIFICIAL FLAVORING,
BHA and BHT,
PROPYL GALLATE.
SODIUM PROPIONATE,
MONO and
DIGLYCERIDES

BOOKS
LAST CHANCE BEFORE EVERYTHING GOES ON MICROFILM
SALE
S. Harris

MADE POSSIBLE BY A GRANT FROM
INC.
FUNDED BY
J-J INDUST INC
BROUGHT TO YOU BY SMEDLY & CO.

"DON'T MIND ASHLEY. AFTER LOOKING THROUGH A MICROSCOPE ALL DAY, ANYTHING LARGE STARTLES HIM."

INSTITUTE
OF
INSTITUTE
INSTITUTE
FOR
ARTIF
INSTITUTE
FOR
ARTIFICIAL
INTELLIGENCE
S. Harris

"DR. HODGES, HERE, IS FROM ENGLAND, AND HE'S BEEN OBSERVING US FOR 14 YEARS. MR. FERRELL, AN AMERICAN, HAS BEEN HERE ONLY THREE WEEKS. MONIQUE CORVEAU, FROM PARIS, HAS PRACTICALLY BEEN LIVING WITH US FOR ABOUT NINE YEARS..."

I hope the Nobel committee hears about this.
Where are the TV cameras?
This could increase my fee 25%.
I'll probably get three chapters for my book out of this.
S. Harris

"I HOPE THEY DON'T ISOLATE ME. THEY'LL NEVER BELIEVE I'M HARMLESS."

1.
2.
3.
4.
DARWIN

CLONING LABS
CLONING LABS
CLONING LABS
S. Harris

"HE HASN'T MADE ANY PROGRESS WITH HIS THEORIES RECENTLY, SO HE'S BEEN WORKING ON HIS RESEMBLANCE TO EINSTEIN."

8×7=56
$$j = j_0 \sin\left(\phi_2 - \phi_1 - \frac{2e}{\hbar}\int_1^2 A\,dx\right).$$
$$\frac{\partial}{\partial t}\left(\phi_2 - \phi_1 - \frac{2e}{\hbar}\int_1^2 A\,dx\right)$$
$$= -\frac{1}{\hbar}(\mu_2 - \mu_1) = \frac{2eV}{h}.$$
S. Harris

AFTER ANTOINE LAVOISIER DISCOVERED OXYGEN, HIS WHOLE FAMILY WOULD BREATHE IT REGULARLY AS A SHOW OF SUPPORT

EARLY SCIENTIFIC FRAUD:
YOUNG THOMAS EDISON TRIED TO PASS OFF A CONTAINER FILLED WITH FIREFLIES AS AN INCANDESCENT BULB

PERSONAL AIR BAG

"IT STARTED WITH A SIMPLE CASE OF PEER-REVIEW."

IF NEWTON WAS A BOTANIST...

Ah, ha— the stem of an apple is not permanently fixed to the tree.

S. Harris

"OUR REPUTATION FOR LONGEVITY IS BASED ON SEVERAL FACTORS: HARD WORK, SIMPLE FOOD, LACK OF STRESS, AND THE INABILITY TO COUNT CORRECTLY."

GREAT MOMENTS IN SHOPPING

LOUIS PASTEUR BUYING HIS FIRST QUART OF PASTEURIZED MILK

If he continues treatment for another month or so, I should have enough material for my book of case histories.
IF I CONTINUE TREATMENT FOR ANOTHER MONTH OR SO, I SHOULD HAVE ENOUGH MATERIAL FOR MY EXPOSÉ OF PSYCHOANALYSIS.
S. Harris

"WE HOPE TO MAKE ANTIBIOTICS, INTERFERON AND DIAGNOSTIC PRODUCTS, BUT JUST TO BE ON THE SAFE SIDE, WE'RE STARTING OUT WITH A LINE OF SHAMPOOS."

WATER
DESALINATION
PLANT
SALT: 15¢ Lb.
S. Harris

SO MUCH FOR THE QUESTION OF WHETHER OR NOT A CURVEBALL REALLY CURVES.
s.harris

"EYE IRRITATION AND COUGHING? YOU CALL THAT TOXIC?"

"I THOUGHT HE WOULD RUN ALL SORTS OF SCIENTIFIC TESTS."

"ONE NICE THING ABOUT BEING A CEPHALOPOD— YOU PUT ON YOUR SHOES, YOU PUT ON YOUR HAT, AND YOU'RE DRESSED."

$E=mc^2$
S. Harris

"WE DID THE WHOLE ROOM OVER IN FRACTALS."

"THEY DID IT AGAIN — NOT A WORD IN THE WEATHER REPORT ABOUT AN ICE AGE."

"WE CREATE IT, WE CLEAN IT UP—BUSINESS COULDN'T BE BETTER."

UNIFORMLY ACCELERATED MOTION:
$x = \frac{1}{2}at^2 + v_0 t + x_0$
CURVATURE:
$K = d\gamma/ds$
S. Harris

RISKS
BENEFITS
HAMBURGER $2.35
CHEESEBURGER $2.95
TUNA SALAD $2.75
EGG SALAD $2.65
OMELETTE $3.10
BEEF STEW $3.45
S. Harris

"WE'VE DISCOVERED A MASSIVE DUST AND GAS CLOUD WHICH IS EITHER THE BEGINNING OF A NEW STAR OR JUST A HELL OF A LOT OF DUST AND GAS."

OIL SOLIDIFICATION PLANT
COAL LIQUIFICATION PLANT
S. Harris

"I THOUGHT THEY ONLY TESTED <u>DRUGS</u> ON GUINEA PIGS."

THE UNIVERSE BEFORE
THE BIG BANG
(ACTUAL SIZE)

S. Harris